T0132841

FLORA OF TROPICAL EAST AFRICA

DIOSCOREACEAE

E. MILNE-REDHEAD

Twining herbs with annual stems arising from tubers or rhizomes, rarely stems self-supporting. Leaves alternate or opposite, often ovate-cordate, but sometimes with 3–7 digitate leaflets. Flowers bisexual or unisexual, the latter usually dioecious. Perianth segments biseriate, usually united basally. Stamens (3, 4) 6. Ovary inferior, rarely semi-inferior or superior, (1)3-locular. Fruit a dehiscent capsule, samara or berry.

A family of 5 genera, only one of which occurs in tropical Africa.

DIOSCOREA

L., Sp. Pl.: 1032 (1753) & Gen. Pl., ed. 5: 456 (1754); Bak. in F.T.A. 7: 414 (1898); R. Knuth in E.P. IV. 43: 45 (1924)

Twining dioecious pubescent or glabrous herbs. Stems annual, arising from tubers, often prickly especially below. Leaves alternate or opposite, petiolate, often ovate-cordate, usually entire, sometimes deeply or shallowly lobed, occasionally compound with 3–7 leaflets; main nerves of entire leaves conspicuous, digitate, tertiary nervation reticulate. Aerial tubers sometimes occur, arising in leaf-axils. Male inflorescence spicate, racemose or rarely cymose, axillary or forming panicles at the leafless ends of branches. Perianth suberect or spreading. Stamens 6, all fertile, or 3 reduced to staminodes, inserted at base and shorter than the perianth. Female inflorescence spicate, axillary. Perianth similar to ♂, with an oblong triquetrous 3-locular ovary immediately beneath; styles 3, short; ovules 2 per locule. Capsule rigid, deeply 3-lobed or triangular-ellipsoid, dehiscing into 3 valves. Seeds up to 2 per locule, variously winged or rarely wingless.

A mainly tropical genus of about 600 species, occurring in both the Old World and the New, most abundant in tropical Asia.

Several species of *Dioscorea* have been cultivated in East Africa. No. 4, *D. bulbifera*, a native species, has been much ennobled in cultivation since the remote past. No. 5, *C. alata* was introduced to Zanzibar from Asia many centuries ago and, spread to the West Coast by early Portuguese navigators, probably reached Uganda through the Congo forests. No. 17, *D. dumetorum* has been slightly ennobled by cultivation and 10, *D. minutiflora* may have been cultivated in places.

Also cultivated in East Africa, particularly in Uganda, are the two cultivar yams, the yellow Guinea yam, *D. cayenensis* Lam., and the white Guinea yam, *D. rotundata* Poir., now often treated as a variety of *D. cayenensis*, which probably originated by selection and hybridization from various West African species of section *Enantiophyllum*. Certain native species are also eaten in times of famine. For further details of these useful yams see J. M. Dalziel, Useful Pl. W. Trop. Afr.: 488 (1937).

The following sections of the genus are represented in the area covered by this work. The order of sections is that adopted by Burkill (in J.L.S. 56: 402 (1960)). *Borderea* (Miégev.) Benth.—sp. 1; *Macroura* (R. Knuth) Burkill—sp. 2; *Opsophyton* Uline emend. Burkill—spp. 3, 4; *Enantiophyllum* Uline—spp. 5–10; *Asterotricha* Uline—spp. 11–13; *Rhacodophyllum* Uline—sp. 14; *Macrocarpae* Uline—spp. 15, 16; *Lasiophyton* Uline emend. Burkill—spp. 17–19.

KEY TO PLANTS WITH MALE FLOWERS*

1. Stems twining to the left**; leaves simple or
   compound 2
   Stems twining to the right; leaves simple 11
2. Leaves simple (entire or lobed) 3
   Leaves compound 9
3. Perianth ± 10(–13) mm. across when flattened; leaves entire with a slightly
   wavy margin, or shallowly to deeply
   lobed (fig. 3/4–6, p. 8) . . . 14. *D. buchananii*
   Perianth up to 5 mm. across; mature leaves
   entire (juvenile leaves of *D. sansibarensis* often shallowly lobed) 4
4. Lower surface of leaf glabrous or with some
   inconspicuous puberulence at the base 5
   Lower surface of leaf densely or sparsely
   beset with thin wavy adpressed hairs,
   glabrescent in age 8
5. Stamens 3 (with 3 staminodes); dwarf plant
   with stems less than 2 m. long; leaves
   up to 6 cm. long; tubers perennial,
   growing from within, retaining ± same
   shape . . . . . . 1. *D. gillettii*
   Stamens 6; plants and many leaves larger;
   tubers perennial with annual incre-
   ments added as lobes, or replaced
   annually 6
6. Leaves mostly opposite on flowering shoots;
   inflorescences up to 30 cm. long; mature
   leaves heart-shaped, up to 46 cm. long,
   with a thickened acumen ± 5 cm. long;
   juvenile leaves often shallowly lobed
   (fig. 3/1–3, p. 8); tuber perennial . 2. *D. sansibarensis*
   Leaves alternate; inflorescences 5–10(–15)
   cm. long; mature leaves less than 20 cm.
   long, without thickening of the acumen,
   juvenile leaves entire; tuber replaced
   annually 7
7. Flowers directed at right-angles to the axis
   of the inflorescence; perianth spreading flat . . . . . . 3. *D. asteriscus*
   Flowers directed towards apex of inflorescence; perianth not spreading . . 4. *D. bulbifera*
8. Inflorescence racemose, with flowers all single
   at the nodes . . . . . 15. *D. preussii*
   Inflorescence with single pedicellate flowers
   or 2–6-flowered cymules at each node . 16. *D. hylophila*
9. Leaflets 3–5(–7), variable in shape; lateral
   nerves pinnate; flowers hidden by
   bracts in catkins; stamens 3, with 3
   staminodes . . . . . 19. *D. quartiniana*

* Alternative key on p. 5.
** Twining to the left, i.e. sinistrorse (as used here and by Burkill), if the observer
sees the stem mounting to the left; and to the right, i.e. dextrorse, if the observer sees
the stem mounting to the right.

Leaflets 3; lateral nerves arising just above
base and curving towards apex; flowers
not in catkins; stamens 6 10

10. Ultimate spikelets of inflorescence up to 15
mm. long, subsessile or on peduncles
up to 5 mm. long; leaflets up to 21 cm.
long . . . . . . 17. *D. dumetorum*
Ultimate spikelets of inflorescence up to 25
mm. long, on peduncles up to 15 mm.
long; leaflets up to 28 cm. long . . 18. *D. cochleari-apiculata*

11. Plants beset with stellate or dendroid indumentum, or at least with sparse stellate
indumentum on young stems and leaves 12
Plants glabrous 14

12. Stamens 3, with 3 staminodes . . . 12. *D. hirtiflora*
Stamens 6, all fertile 13

13. Leaves heart-shaped, densely covered with
stellate indumentum beneath, even in
age . . . . 11. *D. schimperana*
Leaves ovate with cordate base, sparsely or
rarely densely covered with stellate
indumentum beneath, often glabrescent
in age . . . . 13. *D. longicuspis*

14. Stems 4-winged or angled; flowers on a zigzag axis . . . . 5. *D. alata*
Stems terete; axis of inflorescence straight 15

15. Leaves ovate, elliptic or rarely suborbicular,
with 2 strong secondary nerves running
near the margin . . . . 9. *D. smilacifolia*
Leaves not as above 16

16. Perianth scarious; flowers densely arranged
on the inflorescence, early caducous . 8. *D. baya*
Perianth at least in part petaloid; flowers
not densely arranged 17

17. Stems arising from ends of horizontally
radiating branches of perennial rootstock, prickly; tubers several, vertical
on underside of rootstock branches;
roots not thorny; leaf-blades broadly
ovate, orbicular or the uppermost
reniform; perianth subglobose with a
small basal scarious zone . . . 10. *D. minutiflora*
Stems arising from annually replaced vertical
tubers protected above by horizontal
thorny roots; other characters not
combined 18

18. Leaves ovate, ovate-lanceolate, rarely deltoid, basally widely cordate or subtruncate; perianth longer than wide,
with the lower half scarious . . 6. *D. odoratissima*
Leaves ovate-lanceolate or more usually
deltoid, with the basal corners sometimes auriculate, basally rounded or
obscurely cordate; perianth subglobose,
with a small basal scarious zone . . 7. *D. lecardii*

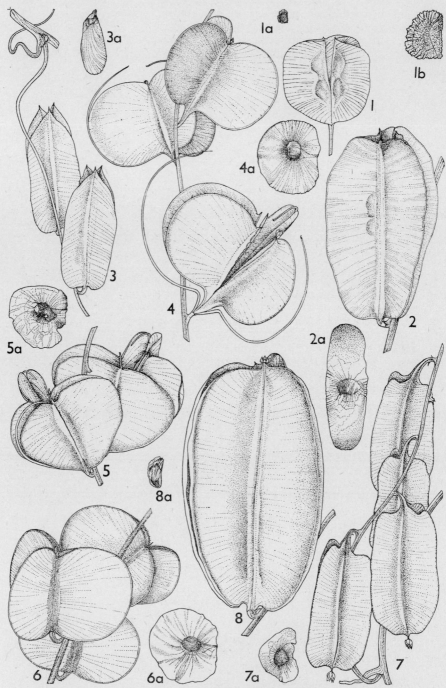

FIG. 1. Capsules (× 1) and seeds (× 1 unless otherwise stated) of species of *Dioscorea*. **1**, *D. gillettii* (**1a**, × ¾; **1b**, × 2½); **2**, *D. sansibarensis*; **3**, *D. astericus*; **4**, *D. odoratissima*; **5**, *D. schimperana*; **6**, *D. hirtiflora*; **7**, *D. buchananii*; **8**, *D. hylophila*. 1, from *Gillett* 13434; 2, from *Faulkner* K. 580 and *Vaughan* 917; 3, from *Batty* 1090; 4, from *Tweedie* 3625, *Bullock* 3149 and *Newbould & Harley* 4309; 5, from *Mutimushi* 2258 and *Verdcourt* 2817; 6, from *Faulkner* K. 670; 7, from *Greenway & Polhill* 11436; 8, from *Robertson* 1060 and *Drummond & Hemsley* 3152. Drawn by Mrs. J. A. Lowe.

KEY TO PLANTS WITH FEMALE FLOWERS OR FRUITS

1. Inflorescence up to 1·5 cm. long, 2-flowered;
   capsule as fig. 1/1 with nerves extending
   from the suture towards the axis; seeds
   without a wing; dwarf plant with stems
   less than 2 m. long . . . .   1. *D. gillettii*
   Inflorescence much longer; capsule without
   such nerves; plants much taller   2
2. Stems twining to the left (see footnote on
   p. 2)   3
   Stems twining to the right; seeds winged all
   round; leaves simple   10
3. Leaves simple (entire or lobed)   4
   Leaves compound; seeds winged at basal end   8

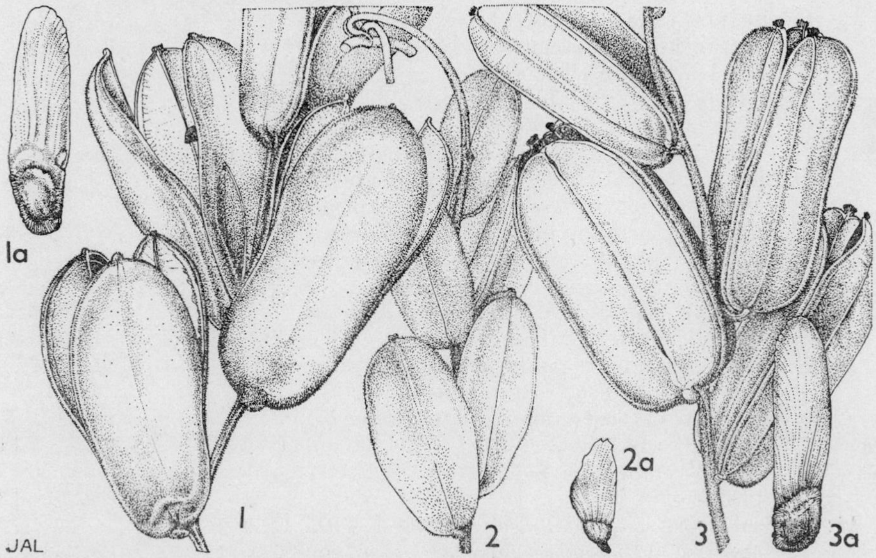

FIG. 2. Capsules and seeds of species of *Dioscorea*. **1**, *D. cochleari-apiculata*, × ⅜; **2**, *D. quartiniana* var. *quartiniana* × 1; **3**, *D. dumetorum*, × 1. 1, from *Lewalle* 3747; 2, from *Peal* 5; 3, from *Archbold* 1133. Drawn by Mrs. J. A. Lowe.

4. Perianth up to 10 mm. across; capsule as
   fig. 1/7; seeds winged at both ends
   with a narrow wing between; leaves
   entire with a slightly wavy margin, or
   shallowly to deeply lobed (fig. 3/4–6, p. 8)   14. *D. buchananii*
   Perianth up to 5 mm. across; mature leaves
   entire (juvenile leaves of *D. sansibarensis*
   often shallowly lobed)   5
5. Lower surface of leaf glabrous or with some
   inconspicuous puescencbe at the base   6
   Lower surface of leaf densely or sparsely
   beset with thin wavy adpressed hairs,
   glabrescent in age; capsule as in fig. 1/8;
   seeds winged at both ends*. . . .   15. *D. preussii*
                                          16. *D. hylophila*

* Two species apparently indistinguishable from ♀ material, but with distinct ♂ inflorescences and distributions.

6. Capsule (fig. 1/2, p. 4) ± 5 cm. long; seeds
winged at both ends; leaves mostly
opposite on flowering shoots; mature
leaves heart-shaped, up to 46 cm. long,
with a thickened acumen ± 5 cm. long;
juvenile leaves often shallowly lobed
(fig. 3/1–3, p. 8); tuber perennial   .   2. *D. sansibarensis*
Capsule (fig. 1/3) 2–2·5(–3) cm. long; seeds
winged at basal end; mature leaves less
than 20 cm. long, without thickening
of acumen; juvenile leaves entire;
tuber replaced annually 7

7. Ovary restricted just below perianth;
perianth-lobes spreading at an angle;
capsule as fig. 1/3 .   .   .   .   3. *D. asteriscus*
Ovary not restricted just below perianth;
perianth-lobes ascending; capsule as
fig. 1/3, but smaller .   .   .   .   4. *D. bulbifera*

8. Leaflets (3–)5–7, variable in shape; lateral
nerves pinnate; capsule as fig. 2/2,
p. 5, glabrescent in age .   .   .   19. *D. quartiniana*
Leaflets 3; lateral nerves arising just above
base and curving towards apex; capsule
usually remaining pubescent even in age 9

9. Leaflets up to 21 cm. long; capsule as
fig. 2/3; pubescent .   .   .   .   17. *D. dumetorum*
Leaflets up to 28 cm. long; capsule as
fig. 2/1, densely pubescent .   .   .   18. *D. cochleari-apiculata*

10. Capsules ascending; plants beset with stellate or dendroid indumentum or at
least with scattered stellate hairs on
young stems and leaves 11
Capsule descending, i.e. pedicel not recurved;
plants glabrous 13

11. Leaves heart-shaped, densely covered with
stellate indumentum beneath, even in
age; capsule as fig. 1/5 .   .   .   11. *D. schimperana*
Leaves usually ovate with a cordate base;
stellate indumentum scattered on lower
surface, sometimes glabrescent in age,
rarely dense 12

12. Stellate indumentum scattered often thickly
on lower surface of leaves; capsule as
fig. 1/6; leaves up to 10 cm. long   .   12. *D. hirtiflora*
Stellate indumentum scattered on lower
surface of leaves, or restricted to the
nerves; capsule as fig. 5/5 (p. 18);
leaves up to 14 cm. long .   .   .   13. *D. longicuspis*

13. Tuber perennial; lobes developing below a
woody horizontal surface; leaf-blade
ovate or reniform; capsule not known .   10. *D. minutiflora*
Tuber replaced annually; capsule as fig. 1/4   .   .   . 14

14. Stems 4-winged or 4-angled   .   .   .   5. *D. alata*
Stems terete 15

15. Leaves ovate, elliptic or rarely suborbicular,
with 2 strong secondary nerves running
near the margin . . . . 9. *D. smilacifolia*
Leaves not as above; tubers protected by
thorny roots 16
16. Leaves ovate, rounded basally; perianth
scarious . . . . . . 8. *D. baya*
Leaves ovate, ovate-lanceolate, or deltoid,
basally widely cordate, rounded or subtruncate, sometimes auriculate; perianth
not scarious 17
17. Leaves ovate, ovate-lanceolate, rarely deltoid, basally widely cordate or subtruncate; capsule glaucous . . 6. *D. odoratissima*
Leaves ovate-lanceolate or more usually
deltoid, with the basal corners sometimes auriculate, basally rounded or
obscurely cordate; capsule not glaucous 7. *D. lecardii*

1. **D. gillettii** *Milne-Redh.* in K.B. 17 : 177 (1963). Type : Kenya, Northern
Frontier Province, Moyale, *Gillett* 14137 (K, holo. !)

Tuber perennial, much compressed with a flat base and an obscurely
lobed edge. Twining stems up to 1·5 m. long, glabrous. Leaves alternate
or rarely subopposite; petiole up to 5 cm. long; blade ovate-cordate,
narrowed to an acute scarcely acuminate apex and with rounded basal lobes,
up to 6 cm. long and 4 cm. broad. Aerial tubers apparently absent. Inflorescences glabrous. Male 1 per leaf-axil, very variable in length, from less
than 1 cm. to 14 cm. long, with single shortly pedicellate flowers or short
2–4-flowered cymules at each node, apically appearing paniculate as the
cymules arise in the axils of bracts; perianth spreading, ± 5 mm. across;
segments elliptic, the inner rather broader. Stamens 3; staminodes 3,
shorter than the stamens, filiform. Female inflorescence 1 per leaf-axil, up
to 1·5 cm. long, 2-flowered; perianth spreading, ± 6 mm. across. Ovary
± 1 cm. long. Capsule as in fig. 1/1, p. 4, with parallel nerves extending
from the suture towards the axis, probably not reflexed. Seeds (fig. 1/1a, b)
without a wing.

KENYA. Northern Frontier Province : Dandu, 6 June 1952 (fr.), *Gillett* 13434 !; Kitui
District : Kandelongwe to Sosoma track, 12 Dec. 1952 (fl.), *L. C. Edwards* 195 ! &
Nzui [Nzue], 19 Jan. 1943 (fr.), *Bally* 1938 !
DISTR. K1, 4; S. Ethiopia (Neghelli)
HAB. *Acacia, Commiphora* scrub; 450–1020 m.

NOTE. This very inconspicuous species has seldom been collected. It is of particular
interest as it belongs to the hitherto monotypic section *Borderea*, the only other
species being the European *D. pyrenaica* Grenier. Both these plants inhabit season-
ally dry inhospitable habitats, but the nature of the habitats is completely different,
for *D. pyrenaica* grows on south-facing alpine screes.

2. **D. sansibarensis** *Pax* in E.J. 15 : 146 (1892); Harms in P.O.A. C : 146
(1895); Bak. in F.T.A. 7 : 418 (1898); V.E. 2 : 359 (1908); R. Knuth in E.P.
IV. 43 : 87 (1924); Burkill in B.J.B.B. 15 : 348 (1939); U.O.P.Z. : 233, fig.
(1949); Burkill & Perrier in Fl. Madag. 44 : 4, fig. 1 (1950); Troupin,
Fl. Sperm. Parc Nat. Garamba 1 : 213 (1956); F.P.S. 3 : 295 (1956); Miège in
F.W.T.A., ed. 2, 3 : 152 (1968); Verdc. & Trump, Common Poisonous Pl.
E. Afr. : 194, fig. 20/a–c (1969). Type : Tanzania, Bagamoyo, *Hildebrandt*
1284 (B, holo.)

Tuber perennial, becoming very large in age, depressed globose, flattened
below, hollowed towards the centre, developing roundish lobes in age, up to

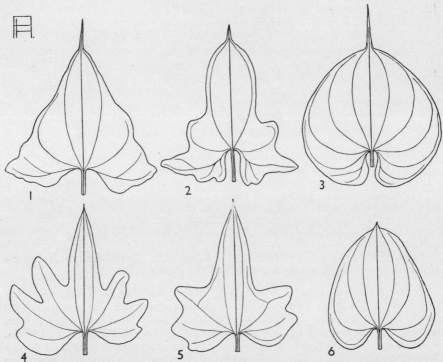

FIG. 3. Variation of leaves in *Dioscorea* species, much reduced, not to scale. *D. SANSIBARENSIS*—1, 2, young leaves; 3, mature leaf. *D. BUCHANANII*—4, 5, young leaves, from same plant; 6, usual form of older leaves. Drawn by Miss P. Halliday.

40 cm. in diameter and 15 cm. deep, the upper surface near ground-level. Plant glabrous. Twining stems 25 m. or more long. Leaves normally opposite; petiole 8–12(–26) cm. long; blade heart-shaped, up to 20(–46) cm. long (including acumen) and 23(–58) cm. broad, with a thickened acute acumen 5(–20) cm. long; towards base of stem leaves replaced by fleshy auricles; leaf-blades of young plants variously shaped and lobed (see fig. 3/1–3). Aerial tubers irregularly roundish, up to 6 cm. diameter, deep purplish. Inflorescences pendulous, spicate. Male 1 or 2 per leaf-axil or forming a terminal leafless panicle; spikes up to 50 cm. long, with 1 or 2 downwardly directed flowers at each node, not opening wide, ± 5 mm. long. Stamens 6. Female inflorescences 1–3 per leaf-axil, up to 48 cm. long; flowers solitary at a node, directed downwards; perianth not opening wide, ± 3 mm. in diameter. Ovary ± 6 mm. long. Capsule as in fig. 1/2, p. 4, ± 5 cm. long and 2·5 cm. in diameter, directed upwards. Seeds (fig. 1/2a) winged at both ends, ± 3·5 cm. long.

UGANDA. Bunyoro District: Rabongo Forest, 24 July 1964 (sterile), *H. E. Brown* 2105
KENYA. Kwale District: Mrima Hill, 31 Aug. 1959 (♂ fl.), *Verdcourt* 2407!; Kilifi
    District: Rabai Hills, *W. E. Taylor* !
TANZANIA. Pangani District: Bushiri, 24 May 1950 (♂ fl.), *Faulkner* K.580!; Iringa
    District: Itundufula, Apr. 1960 (♂ & ♀ fl. & fr.), *Haerdi* 500/0!; Songea District:
    Mbamba Bay, 5 Apr. 1956 (♂ fl.), *Milne-Redhead & Taylor* 9535!; Zanzibar I., Panga
    juu, 10 June 1930 (♂ fl.), *Vaughan* 1342!; Pemba, Fundo I., Nov. 1929 (fr.), *Vaughan*
    917!
DISTR. U2; K7; T3, 6–8; Z; P; Ivory Coast to the Sudan, south to Angola,
    Rhodesia and Mozambique, also in Madagascar
HAB. Lowland rain-forest, riverine forest, *Acacia* woodland, coastal evergreen bush-
    land; often persisting in secondary vegetation; 0–900 m.

SYN.   *D. macroura* Harms in N.B.G.B. 1: 266 (1897); Bak. in F.T.A. 7: 416 (1898);
De Wild., Ann. Mus. Congo, Bot., sér. 5, 2: 22, t. 28/1–5 (1907); V.E. 2:
359, fig. 253/A–E (1908); R. Knuth in E.P. IV. 43: 87, fig. 19/A–E (1924);
F.W.T.A. 2: 382 (1936). Types: Cameroun, Yaoundé, *Zenker* 620 (B, syn.)
& *Zenker & Staudt* 414 (B, syn., K, isosyn.!)
D. welwitschii Rendle in Cat. Afr. Pl. Welw. 2: 39 (1899); R. Knuth in E.P.
IV. 43: 324 (1924). Type: Angola, Golungo Alto, *Welwitsch* 4041 (LISU or
BM, holo., K, iso.!)

NOTE.   The thickened acumen is said to contain nitrifying bacteria, see Notes Roy.
Bot. Gard. Edin. 14: 57–72 (1923).

3. **D. asteriscus** *Burkill* in B.J.B.B. 15: 356 (1939); Verdc. & Trump,
Common Poisonous Pl. E. Afr.: 192, fig. 20/h–j (1969). Type: Malawi,
without locality, *Buchanan* 11 (K, holo.!)

Tuber annual, spherical (*fide* Perdue & Kibuwa). Plant glabrous. Twining stems, up to 3 m. long. Leaves alternate; petiole up to 8 cm. long;
blade heart-shaped, rather gradually narrowed to an acute acumen, up to
14(–18) cm. long and broad, often much smaller. Aerial tubers irregularly
subglobose, up to 5 cm. in diameter, deep purplish. Inflorescences pendulous.
Male racemose, up to 4 per leaf-axil or forming a terminal leafless panicle;
racemes 10(–15) cm. long; flowers standing out on 1–2 mm. long pedicels at
right-angles to the axis; perianth in flower spreading, star-shaped, ± 4 mm.
across; tepals lanceolate, very acute. Stamens 6. Female inflorescences
spicate, 1–3 per leaf-axil, up to 30 cm. long; flowers solitary at a node,
directed downwards; perianth slightly spreading, ± 3(–6) mm. across.
Ovary ± 4 mm. long, restricted just below the perianth. Capsule as in
fig. 1/3, p. 4, 2·5(–3) cm. long, 1·2–1·5 cm. in diameter, directed upwards.
Seeds (fig. 1/3a) winged at basal end, 16 mm. long.

KENYA.   Kericho District: N.W. of Ngoina Tea Estate, 14 Dec. 1967 (♂ fl.), *Perdue &
Kibuwa* 9379!; Kilifi District: Mida, ♂ fl., *R. M. Graham* A520 in *F.D.* 1909! &
Arabuko-Sokoke Forest, Jilori, 26 Nov. 1961 (♀ fl.), *Polhill & Paulo* 860! & Rabai,
Aug. 1937 (young fr.), *V. G. van Someren* in *C.M.* 7164!
TANZANIA.   Tanga District: about 6·5 km. ENE. of Korogwe, 27 June 1953 (♀ fl.),
Drummond & Hemsley 3055!; Mpwapwa, 26 May 1938 (♂ fl. & fr.), *Hornby* 503!;
Morogoro District: Morogoro, Bahati, 21 Apr. 1935 (♂ fl.), *E. M. Bruce* 1083!;
Pemba I., Mvumoni, 19 Aug. 1929 (♂ fl.), *Vaughan* 434A!
DISTR.   ?U2; K5, 7; T1–3 5, 6, 8; ?Z; P; Zaire (Katanga), Zambia, Malawi, Rhodesia,
Mozambique and South West Africa
HAB.   Rain-forest, lowland dry evergreen forest, coastal evergreen bushland in forested
ravines and among rocks in lower rainfall country; 0–1400(–2300) m.

SYN.   [*D. sativa* sensu Bak. in F.T.A. 7: 415 (1898), pro parte *nec* Thunb., *non* L.]
[*D. bulbifera* sensu R. Knuth in E.P. IV. 43: 88 (1924), pro minore parte, *non* L.]

VARIATION.   Two specimens (from Uganda, Ankole, Buganda Forest, Feb. 1939 (♂ fl.),
Cree 254, and Tanzania, Shinyanga, ♂ fl., *Bax* 402) have exceptionally long pedicels,
especially those in the lower part of the inflorescences, and these tend to ascend
rather than spread at right-angles to the axis. It is possible that these specimens
may represent a recognizable infraspecific taxon, but a decision cannot be made
until further material is available for study.

NOTE.   *D. asteriscus* has been much confused with the wild variety of *D. bulbifera*,
which has rather similar rounded aerial tubers; in fact in the sterile state or in fruit
it has not been found possible to distinguish these species. *D. asteriscus* is confined
to the eastern side of Africa, where it has been collected more frequently than has
D. bulbifera. A sterile specimen from Zanzibar I., Chwaka, 5 June 1963, *Faulkner*
3200, is probably *D. asteriscus*.

4. **D. bulbifera** *L.*, Sp. Pl.: 1033 (1753); Harms in P.O.A. C: 146 (1895);
R. Knuth in E.P. IV. 43: 88, fig. 19/F–L (1924); F.W.T.A. 2: 382, fig. 315
(1936); Burkill in B.J.B.B. 15: 357 (1939); Burkill & Perrier in Fl. Madag.

44: 24 (1950); Verdc. & Trump, Common Poisonous Pl. E. Afr.: 196, fig. 20/d (1969). Lectotype: plate facing p. 217 in Hermann, Paradisus Batavus (1698); no specimen in Hermann Herbarium

Tuber perennial, usually irregularly subglobose, but occasionally elongate, and sometimes absent. Plant glabrous or the leaf-blade inconspicuously puberulous at the base beneath. Twining stems up to 12 m. long. Leaves as in 3, *D. asteriscus*. Aerial tubers irregularly subglobose or markedly angular, up to 7 cm. in diameter, brown. Inflorescences as in *D. asteriscus*, except that the ♂ flowers are sessile and directed downwards, towards apex of inflorescence. Perianth lobes of ♂ flower not spreading; tepals lanceolate, up to 2 mm. long. Perianth lobes of ♀ flower directed towards apex of inflorescence, up to 2 mm. long. Capsule and seeds similar to those of *D. asteriscus*, but capsule only 2 cm. long and 1·2 cm. in diameter.

var. bulbifera

Tuber present; aerial tubers rounded, poisonous.

UGANDA. Mubende District: Kakumiro, 1 Oct. 1957 (♂ fl.), *Peal* 11 ! & 27 Sept. 1957 (♀ fl., young fr.), *Peal* 10 !
TANZANIA. Ufipa District: Lake Rukwa, Milepa, 9 Apr. 1934 (sterile), *Michelmore* 1011 !
DISTR. U4; T?1, 4; central and western Africa, tropical Asia and Polynesia
HAB. Lowland rain-forest; 1150–1230 m.

NOTE. *Kadesha* in *Western Research Station* 1442, sterile, from Mwanza District, Ukiriguru, may belong here.

var. anthropophagorum (*A. Chev.*) *Summerh.* in Trans. Linn. Soc., Zool. 19: 293 (1931); Prain & Burkill in Ann. Roy. Bot. Gard. Calc. 14: 117, t. 51/2 (1936); Burkill & Perrier in Fl. Madag. 44: 25 (1950). Type: none cited

Tuber reduced or absent; aerial tubers angular, edible.

UGANDA. Bunyoro District: without locality, *Dawe* 753 !*
TANZANIA. Lushoto District: Amani, Sept. 1949, *Greenway* !; Ufipa District: Lake Rukwa, Milepa, 9 Apr. 1934 (sterile), *Michelmore* 1012 !; Kilosa District: Ilonga, 13 June 1968 (♀ fl., young fr.), *Robertson* 1064a !; Pemba I., Piki, 28 Sept. 1928 (sterile), *Vaughan* 683 !
DISTR. U?2; T2–4, 6; P; central and western Africa
HAB. Lowland rain-forest and secondary evergreen bushland in high rainfall areas, persisting in abandoned cultivated ground and by perennial rivers; 0–480 m.

SYN. *D. anthropophagorum* A. Chev. in Veg. Ut. Afr. Trop. Fr. 8: 357 (1913), as " anthrophagorum "

SYN. (of species as a whole). *D. sativa* Thunb., Fl. Jap.: 151 (1784); Bak. in F.T.A. 7: 415 (1898), *non* L. (1753)
    D. bulbifera L. var. *sativa* Prain, Bengal Pl.: 1066 (1903) & ed. 2: 801 (1963)

NOTE. *D. bulbifera* is the most widespread species of the genus, extending through tropical Africa and Asia to the remotest Pacific islands. It has been in cultivation in Asia and Africa for thousands of years and many ennobled races (cultivars) exist in Asia (see Prain & Burkill in Ann. Roy. Bot. Gard. Calc. 14: 111–132 (1936)). The best-known Asiatic race is known as var. *sativa*, whilst in Africa a different edible race with angular aerial tubers has been evolved, known as var. *anthropophagorum*. This has been introduced to tropical America.
    Unfortunately in the absence of knowledge of the aerial tubers, and of the presence or absence of terrestrial tubers, it is not possible to distinguish the cultivated variety from the typical wild variety (var. *bulbifera*) and many gatherings in herbaria cannot be satisfactorily determined. As a result the distributions given above are very incomplete. An added complication is that, in the sterile state, it is not possible to distinguish var. *bulbifera* from the closely allied *D. asteriscus* Burkill, which is apparently more frequent than *D. bulbifera* in eastern Africa.

* Probably this variety, as Dawe had named it *D. sativa*.

5. **D. alata** *L.*, Sp. Pl.: 1033 (1753); Bak. in F.T.A. 7: 417 (1898); R. Knuth in E.P. IV. 43: 265 (1924); F.W.T.A. 2: 382 (1936); Burkill in B.J.B.B. 15: 380 (1939) & in Fl. Males., ser. 1, 4: 330 (1951); Miège in F.W.T.A., ed. 2, 3: 152 (1968); Verdc. & Trump, Common Poisonous Pl. E. Afr.: 192 (1969). Types: Ceylon, *Hermann* 2: 23 (BM, syn.)* & without data, *Linnean Herbarium* 1184.2 (LINN, syn., IDC microfiche!)

Tuber replaced annually, normally cylindric, up to 6 cm. in diameter, descending vertically, but in some cultivars very diversely shaped, branched or expanded above, sometimes with lobes curved or spreading horizontally. Twining stems 4-winged or -angled, up to 10 m. long, glabrous. Leaves opposite or the lower alternate, glabrous; petiole up to 9(–10) cm. long; blade ovate with widely cordate almost hastate base, apically narrowed to an acute acumen, up to 12(–13) cm. long and 8(–10) cm. broad. Aerial tubers subglobose or irregularly and narrowly ovoid, up to 12 cm. long. Inflorescences glabrous. Male ± 2 in the leaf-axils or forming axillary terminal panicles in the axils of bracts, spreading; axis zigzag, with the sessile flowers directed forwards and outwards; perianth subglobose, not opening widely, ± 1·5 mm. across. Female 1 per leaf-axil, up to 21(–35) cm. long; perianth triangular-subglobose, ± 5 mm. across. Ovary glabrous. Capsule as in fig. 1/4, p. 4, up to 3·5 cm. in diameter, glabrous. Seeds winged all round.

UGANDA. Mbale District: N. Bugisu, Bumasifwa, 26 Sept. 1958 (sterile), *Peal* 23!; Masaka District: 16 km. from Masaka on road to Kampala, 20 Sept. 1958 (sterile), *Peal* 34!; Mubende District: about 15 km. from Kakumiro, 30 Sept. 1957 (sterile), *Peal* 14!
TANZANIA. Lushoto District: Amani, July 1939 (♂ & ♀ fl.), *Greenway* 5884! & 5885!; Zanzibar I., Kizimkazi, 27 May 1960 (♀ fl.), *Faulkner* 2573! & Makunduchi, 23 June 1961 (♀ fl.), *Faulkner* 2859!
DISTR. U3, 4; T1, 3, 8; Z; tropical eastern Asia, now spread around the humid tropics of both Old and New World
HAB. Cultivated for its edible tuber, and often persisting in secondary forest and bushland; the cited specimens from Uganda are all cultivated, as are those from Amani, but *D. alata* is likely to persist in these areas, as it is doing in Zanzibar; 0–1350 m.

SYN. *D. colocasiifolia* Pax in E.J. 15: 145 (1892), in part, excluding inflorescence. Type: Cameroun, cultivated at Victoria, *Bucholz* (B, holo., K, photo.!)
    D. sapinii De Wild. in Ann. Mus. Congo, Bot., sér. 5, 3: 368 (1912). Type: Zaire, Katanga, Katola, *Sapin* (BR, holo.)

VARIATION. Whilst the above-ground parts of this yam show comparatively little variation, the tuber, largely as a result of long cultivation and selection by man, has developed a number of very different shapes.

NOTE. *D. alata* has been introduced into Africa from the East, and no doubt a number of different cultivars have become established here and there as crop plants. As it is propagated vegetatively, it is usual to find plants of one sex only in any one locality. Mrs. Faulkner searched for but failed to find a male plant in Zanzibar, where it has become established in natural vegetation. The only plants collected by Peal in Uganda were either female or sterile. I have not yet seen a male flowering specimen from East Africa. As plants of the two sexes seldom grow together, the production of fruit is very rare in Africa. However, *D. alata* has become well established in second-ary vegetation, where it has the appearance of a native plant.

6. **D. odoratissima** *Pax* in E.J. 15: 146 (1892); Burkill in B.J.B.B. 15: 381 (1939); Lawton in Journ. W. Afr. Sc. Assoc. 12: 7, fig. 3b (1967). Types: Togo, Bismarckbourg, *Büttner* 103 (B, syn.) & 104 (B, lecto., K, isolecto.!)

* The specimen consists of two leaves only.

Tuber replaced annually, up to 5 cm. in diameter and more than 6 dm. long, descending vertically, protected by horizontal thorny roots above. Twining stems very prickly below, less so above, the lower 3 m. leafless but with thick fleshy stipules, up to 12 m. long, glabrous. Leaves opposite or occasionally alternate, glabrous; petiole up to 5(–7) cm. long; blade ovate, ovate-lanceolate or rarely deltoid, widely cordate or subtruncate at the base, acutely acuminate, up to 10(–15) cm. long and 5(–7·5) cm. broad. Aerial tubers absent. Inflorescences glabrous. Male spicate, ascending, up to 5(–7) cm. long, with the flowers not densely arranged, arising up to 5(–7) in the axils of leaves or at leafless nodes on terminal or lateral branches up to 10(–17) cm. long; perianth longer than wide with the lower half somewhat scarious, greyish or straw-coloured and the upper half petaloid, ± 2 mm. long and 1·5 mm. in diameter apically, narrower below. Female pendulous, 1 per leaf-axil, up to 15(–17) cm. long; perianth subglobose, somewhat flattened, scarcely 2 mm. across. Capsule as fig. 1/4, p. 4, up to 3·8 cm. in diameter, glaucous. Seeds (fig. 1/4a) winged all round.

UGANDA. Bunyoro District: without locality, ♂ fl., *Dawe* 825!; Mbale District: Buluganya, 30 Sept. 1958 (♂ fl.), *Peal* 30!; Mubende District: Kakumiro, about 1 km. along Mubende road, 27 Sept. 1957 (♂ fl.), *Peal* 2!
KENYA. N. Kavirondo District: Malaba Forest, Sept. 1964 (♂ fl.), *Tweedie* 2901! & Aug. 1965 (♀ fl.), *Tweedie* 3086! & Mar. 1969 (fr.), *Tweedie* 3625! & Kakamega, 16 Sept. 1949 (♂ fl.), *Maas Geesteranus* 6235!
TANZANIA. Buha District: Gombe Stream Reserve, Rutanga valley, 21 Jan. 1964 (♂ fl.), *Pirozynski* 258!; Ufipa District; Milepa, Apr. 1937 (♂ fl.), *Lea*!; Songea District: E. of Kitai near R. Rovuma, 17 Apr. 1956 (fr.), *Milne-Redhead & Taylor* 9753!
DISTR. U2–4; K5; T4, 7, 8; widely spread to Sierra Leone in the west and Malawi, Zambia and Angola in the south
HAB. Rain-forest and riverine forest, termite-hills and woodland river-banks; 800–1800 m.

SYN. [*D. praehensilis* sensu Bak. in F.T.A. 7: 418 (1898); Hutch., F.W.T.A. 2: 382 (1936), pro parte; Miège in F.W.T.A., ed. 2, 3: 153 (1968), pro parte; U.K.W.F.: 722 (1974), *non* Benth.]
   D. liebrechtsiana De Wild. in Bull. Herb. Boiss., sér. 2, 1: 53 (1900); R. Knuth in E.P. IV. 43; 297 (1924); Burkill in B.J.B.B. 15: 381 (1939); Miège in F.W.T.A., ed. 2, 3: 154 (1968). Types: Zaire, Kisantu, *Gillet* 591 & 757 (BR, syn.)
   [*D. rotundata* sensu R. Knuth in E.P. IV. 143: 300 (1924), *non* Poir. (1813)]

NOTE. *D. praehensilis* Benth. as recognized by me is a West African species which has no spiny roots protecting its tubers, and has male flowers subglobose in shape and petaloid throughout. I have seen no such plant from East Africa.

7. D. lecardii *De Wild.* in Ann. Mus. Congo, Bot., sér. 5, 1: 19 (1903); R. Knuth in E.P. IV. 43: 295 (1924); Burkill in B.J.B.B. 15: 387 (1939); Miège in F.W.T.A., ed. 2, 3 : 154 (1968). Types: Senegal, without locality, *Lécard* 214 & 235 (both BR, syn.)

Tuber replaced annually, descending vertically, protected by thorny roots. Twining stems unarmed, up to 7·5 m. high, glabrous. Leaves opposite or occasionally alternate, glabrous; petiole up to 2·5(–4) cm. long; blade ovate-lanceolate or more usually deltoid, with the basal corners sometimes auriculate, rounded or very obscurely cordate at the base, tapering to an acute apex, up to 9(–11) cm. long, 2·5(–5) cm. broad; lower leaves probably proportionally broader. Aerial tubers not known, probably absent. Inflorescence glabrous. Male as 6, *D. odoratissima*; perianth subglobose, wholly petaloid or a small basal zone scarious, ± 2 mm. in diameter. Female as 6, *D. odoratissima*. Capsule as *D. odoratissima*, up to 4 cm. in diameter, not glaucous. Seeds winged all round.

UGANDA. Bunyoro District: Biso [Bisu], June 1935 (♂ fl.), *Eggeling* 2060 in *F.D.*
1733!; Teso District: Serere, May 1932 (♂ & ♀ fl.), *Chandler* 678!; Mengo District:
Jumba, July/Aug. 1916 (fr.), *Dummer* 2950!
TANZANIA. Moshi District: Marangu, Apr. 1893 (♂ fl.), *Volkens* 258!
DISTR. U1–4; T2; Senegal E. to Zaire and the Sudan
HAB. Wooded grassland, at edge of thickets and on rocky outcrops, sometimes climbing
among tall grasses; 1050–1800 m.

SYN. *D. mildbraedii* R. Knuth in E.P. IV. 43: 295 (1924). Type: Cameroun, Baja
Highlands, *Mildbraed* 9770 (B, holo.)

NOTE. I follow Burkill in recognizing *D. lecardii* as a species of the wooded grasslands
bordering the rain-forest. It is close to *D. abyssinica* Kunth and to *D. praehensilis*
Benth., species which I do not find in East Africa. From both of these it differs in
the proportionally narrower leaf, usually of a more firm texture.

8. **D. baya** *De Wild.* in Ann. Mus. Congo, Bot., sér. 5, 3: 357, t. 52 (1912);
R. Knuth in E.P. IV. 43: 301 (1924); Burkill in B.J.B.B. 15: 391 (1939).
Type: Zaire, by R. Congo at Bomane, *Claessens* 695 (BR, holo.)

Tuber probably replaced annually, shape not recorded, protected by
thorny roots. Stems with prickles below, up to 27 m. high. Leaves opposite
or alternate, glabrous; petiole up to 6 cm. long; blade ovate or elliptic,
rounded at the base, broadly and shortly acuminate, 3–5-nerved with the
2 primary nerves not running near the margin, up to 6(–9) cm. long and
4·5(–6) cm. broad. Aerial tubers not observed. Inflorescences glabrous.
Male spicate up to 4 cm. long with the flowers usually densely arranged and
early caducous, arising up to 7 in the axils of leaves or more rarely at leafless
node on terminal or lateral branches up to 15 cm. long; perianth subglobose, scarious, ± 1 mm. in diameter. Female 1 per leaf-axil, up to 17 cm.
long; perianth semi-globose, ± 1·5 mm. in diameter; tepals coriaceous.
Ovary glabrous. Capsule as 6, *D. odoratissima*, up to 3·8 cm. in diameter.
Seeds winged all round.

UGANDA. Mengo District: Kipayo, June 1914 (♂ fl.), *Dummer* 859! & Nambigirwa
Forest near Entebbe, Jan. 1932 (♂ fl.), *Eggeling* 372! & Kajansi Forest, 16 km. from
Kampala on Entebbe road, Feb. 1938 (♂ fl.), *Chandler* 2173!
DISTR. U4; Zaire, Gabon, Angola (Zaire) & Zambia (Mwinilunga)
HAB. Lowland rain-forest; 1150–1200 m.

NOTE. Like No. 9, *D. smilacifolia*, this species is on the edge of its recognized range in
Uganda and is clearly very local. The female plant has not been seen in Uganda.

9. **D. smilacifolia** *De Wild.* & *Th. Dur.* in Ann. Mus. Congo, Bot., sér. 2,
1: 58 (1899) & sér. 3, 1: 239 (1901); R. Knuth in E.P. IV. 43: 303 (1924);
F.W.T.A. 2: 382 (1936); Burkill in B.J.B.B. 15: 391 (1939); Miège in
F.W.T.A., ed. 2, 3: 153 (1968). Type: Zaire, Kisangani, Stanley Falls,
Dewèvre 1161 (BR, holo.)

Tuber probably replaced annually, shape and armament not recorded.
Twining stems with scattered prickles below. Leaves opposite or alternate,
glabrous; petiole up to 7 cm. long; blade ovate, elliptic or rarely suborbicular, rounded at the base, broadly and shortly acuminate, conspicuously
3-nerved with the lateral nerves running near the margin, up to 9(–13) cm.
long and 6(–10) cm. broad. Aerial tubers not observed. Inflorescences
glabrous. Male spicate, up to 7 cm. long, arising up to 6 in the axils of
leaves or leafless nodes on terminal or lateral branches up to 40 cm. long;
perianth subglobose, ± 1·5 mm. in diameter. Female 1 per leaf-axil, up to
40 cm. long, often much shorter; perianth semi-globose, ± 1·5 mm. in diameter. Ovary glabrous. Capsule as 6, *D. odoratissima*, up to 3·5 cm. in
diameter. Seeds winged all round.

UGANDA. Mengo District: Mabira Forest, Mulange, Jan. 1920 (♂ fl.), *Dummer* 4364!
& Nabugulo Forest, 17 Dec. [?1918], sterile, *Dummer* 3281!
DISTR. U4; Zaire, Angola (Zaire), Cameroun, Fernando Po and westwards to Sierra
Leone
HAB. Lowland rain-forest; 1200 m.

NOTE. Although yams have been much studied by Peal in Uganda, this species which
is on the edge of its range in Uganda, and said by Dummer to be rare, has been
collected but twice, the female plant never having been recorded within Uganda.

10. **D. minutiflora** *Engl.* in E.J. 7: 332 (1886); Pax in E.J. 15: 146, t. 8
(1892); R. Knuth in E.P. IV. 43: 300 (1924); F.W.T.A. 2: 382 (1936);
Burkill in B.J.B.B. 15: 389 (1939) & in Proc. Linn. Soc. 159: 77 (1947);
Miège in F.W.T.A., ed. 2, 3: 153 (1968). Type: Cameroun, Mungo, *Bucholz*
(B, holo.)

Rootstock perennial, in age consisting of several horizontally radiating
woody arms with a shoot at the outside end, eventually dying back towards
the centre to form separate plants; each arm protecting several vertical
fleshy tubers about 25 cm. long. Twining stems prickly, up to 10 m. high,
glabrous. Leaves opposite or occasionally alternate, glabrous; petiole up
to 10 cm. long; blade broadly ovate, orbicular or the uppermost reniform,
cordate or rounded at the base, shortly acuminate, the lower up to 12 cm.
long and 9 cm. broad, the upper 5 cm. long and 6 cm. broad. Aerial tubers
absent. Inflorescence glabrous. Male as 7, *D. lecardii*. Female as 6, *D.
odoratissima*. Capsule and seeds not seen.

UGANDA. Kigezi District: Kinkizi, Amahenge, Mar. 1946 (♂ fl.), *Purseglove* 2018!;
Mengo District: 13 km. from Kampala along Masaka road, Sept. 1937 (unripe fr.),
Chandler 1937! & Kyagwe, Dec. 1959 (♀ fl.), *Peal & Mukasa* 40!
DISTR. U2–4; Senegal and Guinée east to Zaire and south to Gabon and Angola
(Cabinda)
HAB. Rain-forest; 1050–1750 m.

SYN. *D. praehensilis* Benth. var. *minutiflora* (Engl.) Bak. in F.T.A. 7: 418 (1898)

NOTE. I am not convinced that *D. minutiflora* occurs in a wild state farther east than
Uganda. A specimen from T3 (Amani) so named by Burkill and one from T1 (Bukoba)
are most probably a cultivar, possibly of *D. rotundata* Poir. A cultivated specimen
of *D. minutiflora* is from Lower Kabete, 14 Mar. 1970 (♂ fl.), *Kimani* 219! Another
specimen so named from Kenya (Chyulu foothills, 18 May 1938, *Bally* 652 (*C.M.* 7898))
is sterile and, in my opinion, indeterminable; good specimens of it should be sought
in that area.
　　Two sheets of *D. sublignosa* R. Knuth (in E.P. IV. 43: 304 (1924)) based on a
badly prepared fruiting specimen from T3, Tanga District, Muheza, collected in
Nov. 1907 by *Braun* (*Herb. Amani* 1563) have been examined. The Amani sheet,
now in EA, had been annotated by Burkill as " exactly the plant *Culwick* 20 " (the
Bukoba plant mentioned above), and he names it *D. minutiflora* Engl. The Berlin
holotype is even less satisfactory, but has some fruits in a packet. It has not been
annotated by Burkill. I definitely exclude *D. sublignosa* from the synonymy of
D. minutiflora and consider it best treated as an insufficiently known taxon, probably
a cultivar.

11. **D. schimperana** *Kunth*, Enum. Pl. 5: 339 (1850), as " *schimperiana* ";
A. Rich., Tent. Fl. Abyss. 2: 317 (1851); Th. Dur. & Schinz, Consp. Fl.
Afr. 5: 275 (1893); Bak. in F.T.A. 7: 419 (1898); R. Knuth in E.P. IV.
43: 255 (1924); Burkill in B.J.B.B. 15: 371 (1939); Miège in F.W.T.A.,
ed. 2, 3: 152 (1968); Verdc. & Trump, Common Poisonous Pl. E. Afr.:
196, fig. 20/e–g (1969). Type: Ethiopia, Tigre, Djeladjeranne, *Schimper*
(sect. 3) 1642 (B, holo., K, iso.!)

Tuber replaced annually, cylindric, descending vertically, up to 75 cm. long, branched for last 15 cm., 4·5 cm. in diameter. Twining or scrambling stems up to 8 m. long, densely adpressed stellate-pubescent, glabrescent in age, rarely beset with small prickles below. Leaves usually opposite, occasionally alternate; petiole up to 10(–14) cm. long; blade heart-shaped with an acute thickened acumen, up to 14 cm. long and broad, often much smaller and sometimes proportionally narrower, covered with dense greyish stellate hair beneath, with scattered stellate hairs and glabrescent above. Aerial tubers sometimes present in the axils of the upper leaves, subglobose, ± 1·5(–2·5) cm. in diameter. Inflorescences descending, spicate. Male up to 5(–8) per leaf-axil, or at leafless nodes at the ends of the branches, up to 10(–20) cm. long; flowers directed towards apex of spike; perianth cup-shaped in flower, ± 2 mm. across; segments ovate-lanceolate, obtuse, all stellate-pubescent outside or the inner ones glabrous. Stamens 6. Female up to 4 per leaf-axil, up to 16(–20) cm. long; perianth slightly incurved, ± 1·5 mm. across. Ovary densely stellate-pubescent. Capsule as in fig. 1/5, p. 4, 2·5–3·3 cm. in diameter, glabrescent but remaining stellate-pubescent near the axis, directed upwards. Seeds (fig. 1/5a) winged all round, 1·5–2·0 cm. in diameter.

Uganda. W. Nile District: Mt. Wati, 19 July 1953 (♂ fl.), *Chancellor* 9!; Bunyoro District: Waki R. valley, May 1943 (♂ fl.), *Purseglove* 1578!; Mubende District: about 1 km. from Kakumiro on Mubende road, 27 Sept. 1957 (♀ & ♂ fl.), *Peal* 3! & 4!

Kenya. Turkana District: Kacheliba escarpment, May 1932 (♂ buds), *Napier* 1999!; Trans-Nzoia District: SE. Elgon foothills, June 1964 (♀ & ♂ fl.), *Tweedie* 2845! & 2847!; N. Kavirondo District: Kitosh, Kuywa to Kibabii, 25 July 1951 (♂ fl.), *Greenway & Doughty* 8529!

Tanzania. Ngara District: Nyamyaga [Nyamiaga], Mukagezi, 4 Jan. 1961 (♂ fl.), *Tanner* 5583B!; Iringa District: Mufindi, Nyalawa R., 21 Mar. 1962 (♂ & ♀ fl.), *Polhill & Paulo* 1826! & 1826A!; Songea District: SW. of Kitai near R. Mkaku, 9 Mar. 1956 (♂ & ♀ fl.), *Milne-Redhead & Taylor* 9128! & 9129!

Distr. U1–4; K2, 3, 5; T1, 4, 6–8; Nigeria (Jos Plateau), Cameroun, east to Ethiopia and south to Mozambique and Rhodesia

Hab. Edges and clearings in upland rain-forest, riverine forest, woodland, grassland, often on termite-hills, in ravines and on mountain-tops; 780–2100 m.

Syn. *D. schimperana* Kunth var. *vestita* Pax in E.J. 15: 148 (1892); Th. Dur. & Schinz, Consp. Fl. Afr. 5: 276 (1893); Harms in P.O.A. C: 146 (1895); R. Knuth in E.P. IV. 43: 255 (1924); Burkill in B.J.B.B. 15: 373 (1939). Types: Sudan, Equatoria, Gumango Hill, *Schweinfurth* 3920 (B, syn., K, isosyn.!) & Malawi, Shire Highlands, *Buchanan* 112 (B, syn., K, isosyn.!)
   D. fulvida Stapf in J.L.S. 37: 530 (1906). Type: Uganda, Bunyoro District, *Dawe* 908 (K, holo.!)
   D. hockii De Wild. in B.J.B.B. 3: 277 (1911) & in Ann. Mus. Congo, Bot., sér. 5, 3: 361, t. 64 (1912); R. Knuth in E.P. IV. 43: 256 (1924). Type: Zaire, Katanga, R. Luembe, *Hock* (BR, holo.)
   D. stellato-pilosa De Wild. in Ann. Mus. Congo, Bot., sér. 5, 3: 369, t. 67 (1912); R. Knuth in E.P. IV. 43: 256 (1924). Type: Zaire, Katanga, Katola, *Sapin* 1908 (BR, holo.)

Variation. Considerable variation is shown within this species, chiefly in the degree of indumentum on the leaves, the colour of the indumentum on the inflorescences and the shape of the fruits. These characters grade gradually from one extreme to the other, and show no marked correlation, nor do they appear to be of ecological or phytogeographic importance.

Note. The distribution of this species is of interest, in that the plant occurs on the lower slopes of Elgon, but is absent from the other East African mountains; it keeps largely to the west of the Rift Valley until it crosses to the Southern Highlands of Tanzania, reaching Mahenge in Ulanga District in the east. Var. *nigrescens* R. Knuth, also included under *D. schimperana* by Burkill, is here referred to No. 13, *D. longicuspis* R. Knuth, a local species with a more easterly montane distribution (see p. 18).

12. **D. hirtiflora** *Benth.* in Hook., Niger Fl.: 537 (1849); Bak. in F.T.A. 7: 416 (1898); R. Knuth in E.P. IV. 43: 307, fig. 57 (1924); Burkill in B.J.B.B. 15: 374 (1939); Lawton in Journ. W. Afr. Sci. Ass. 12: 6, t. 4/a (1967); Miège in F.W.T.A., ed. 2, 3: 152 (1968). Type: Nigeria, R. Niger [Quorra], *Vogel* (K, holo.!)

Tuber replaced annually, cylindric, descending vertically, branching at the top (*fide* Lawton), up to 5 cm. in diameter but often very much less. Twining or scrambling stems up to 8 m. long, densely or sparsely beset with spreading or felted stellate or dendroid hairs, rarely (subsp. *orientalis*) almost glabrous, glabrescent in age. Leaves usually opposite, occasionally alternate; petiole up to 5(–9) cm. long; blade ovate-cordate to broadly heart-shaped, acutely acuminate or gradually narrowed to an acute apex, up to 10 cm. long and 9·5 cm. broad, sometimes proportionally wider, more often proportionally narrower, with rather densely scattered stellate hairs beneath, with sparsely scattered stellate hairs and glabrescent above. Inflorescences descending, spicate or racemose. Male up to 3(–5) per leaf-axil, up to 12(–15) cm. long; flowers subsessile or shortly pedicellate, directed towards apex of inflorescence; perianth erect in flower, less than 2 mm. across; segments oblong-lanceolate, obtuse, glabrous or the 3 outer stellate-pubescent outside, up to 3·5 mm. long. Stamens 3; staminodes 3. Female 2(–3) per leaf-axil, up to 16(–23) cm. long; perianth erect, with apices of the tepals incurved, ± 1 mm. across. Ovary densely stellate-pubescent. Capsule as in fig. 1/6, p. 4, 2·5–3.5(–4) cm. diameter, glabrescent but remaining stellate-pubescent near the axis, directed upwards. Seeds (fig. 1/6a) winged all round, up to 2 cm. in diameter.

subsp. **pedicellata** *Milne-Redh.* in K.B. 26: 573 (1972). Type: Zambia, Chingola, *Fanshawe* 2661 (K, holo.!, EA, iso.)

Stems pubescent with stellate or spreading dendroid indumentum, glabrescent in age; leaves normally considerably longer than wide; ♂ flowers shortly pedicellate with outer segments stellate-pubescent.

UGANDA. Mengo District: about 21 km. from Kampala on Entebbe road, Oct. 1931 (fr.), *Hansford* in *Snowden* 2331!
TANZANIA. Kigoma District: Gombe stream, 27 Jan. 1964 (♂ fl.), *Pirozynski* 318!; Ufipa District: Kasanga, *Richards*!; Iringa District, Great Ruaha R. 12 km. W. of Kidatu Bridge, 3 Sept. 1970 (fr.), *Thulin & Mhoro* 839!
DISTR. U4; T4, ?7; Katanga province of Zaire, Zambia, Rhodesia, Malawi (Nyika Plateau)
HAB. Lowland rain-forest, riverine forest and secondary scattered-tree grassland; 770–1200 m.

NOTE. In the absence of ♂ flowers the specimen from Iringa District cited above cannot be confirmed as subsp. *pedicellata*.

subsp. **orientalis** *Milne-Redh.* in K.B. 26: 574 (1972). Type: Tanzania, Handeni District, 16 km. from Korogwe on Handeni road, *Faulkner* 4238A (K, holo.!)

Stems glabrous or with a few stellate hairs when young, soon glabrescent; leaves normally considerably longer than wide; ♂ flowers subsessile with segments all glabrous or the 3 outer with a few stellate hairs basally.

KENYA. Kilifi District: Arabuko, ♂ fl., *R. M. Graham* B512 in *F.D.* 1912*! & Kilifi, 31 July 1936 (♂ fl.), *Moggridge* in *Herb. Amani* H54/36!
TANZANIA. Tanga District: Steinbruch Forest, 29 May 1957 (♂ fl.), *Faulkner* 1982!; Morogoro District: Kingolwira, Mkumbe Hill, July 1935 (fr.), *B. D. Burtt* 5166!; Rufiji District: Mafia I., 25 Mar. 1933 (♂ fl.), *Wallace* 719!
DISTR. K7; T3, 6, 8; Mozambique and Malawi
HAB. Dry evergreen forest, thickets and rocky places; 0–570 m.

* Possibly 2912, as typed label is indistinct.

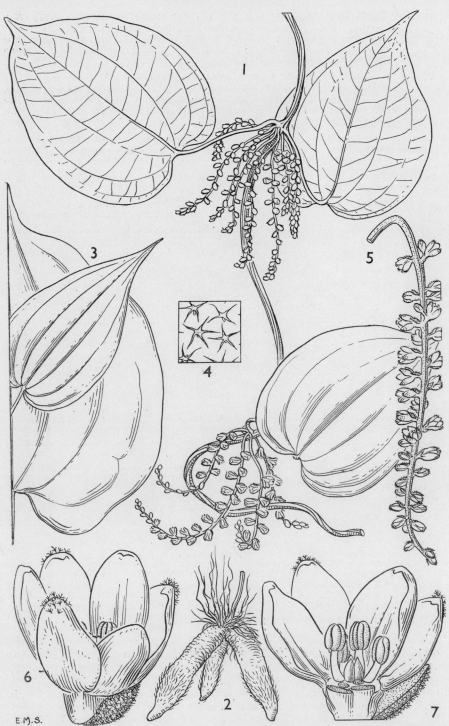

E.M.S.

FIG. 4. *DIOSCOREA LONGICUSPIS* (♂)—**1**, male flowering branch, × 1; **2**, tubers, × $\frac{1}{15}$; **3**, leaves, × $\frac{2}{3}$; **4**, stellate hairs from tip of outer perianth segment, × 32; **5**, part of male inflorescence, × 2; **6**, male flower, × 20; **7**, section of same, × 20. All from *Eggeling* 6766. Drawn by Miss E. M. Stones.

SYN.   *D. lindiensis* R. Knuth in N.B.G.B. 12: 703 (1935). Type: Tanzania, Lindi District, Rondo [Muera] Plateau, *Schlieben* 5971 ! (B, holo. !, K, iso. !)
  D. hirtiflora Benth. var. *grahamii* Burkill in B.J.B.B. 15: 375 (1939), *sine descr. lat., nom. illegit.* Type: Kenya, Kilifi District, Arabuko, *R. M. Graham* B512 in *F.D.* 1912 (K, holo. !)

VARIATION.   *Wallace* 719 from Mafia I. differs in having the ♂ flowers very densely arranged on the axis of the inflorescences, but in other respect agrees with subsp. *orientalis.   Faulkner* K.670 (Pangani District, Bushiri, 1 June 1950) has capsules considerably larger than usual and may be a hybrid with No. 13, *D. longicuspis,* a species not recorded from such a low altitude.

NOTE.   *D. hirtiflora* Benth. subsp. *hirtiflora,* with young stems beset with short sometimes felted stellate indumentum, leaves normally as wide as long, and ♂ flowers sessile with stellate-pubescent outer tepals, occurs widely in forests from Guinée to Cameroun, and south to Angola.

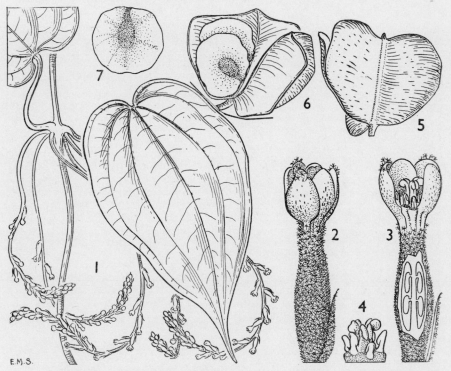

E.M.S.

IG. 5.   *DIOSCOREA LONGICUSPIS* (♀)—**1,** female flowering branch, × 1; **2,** female flower, × 10; **3,** section of same, × 10; **4,** detail of staminodes and styles, × 15; **5,** capsule, × 1; **6,** capsule opened to show seeds, × 1; **7,** seed, × 1. 1, from *Eggeling* 6770C; 2–4, from *Eggeling* 6770B; 5–7, from *Eggeling* 6771. Drawn by Miss E. M. Stones.

13. D. longicuspis *R. Knuth* in N.B.G.B. 11: 1059 (1934). Type: Tanzania, NW. Uluguru Mts., *Schlieben* 3139 (B, holo. !)

Tuber replaced annually, branching at top, the branches cylindric, descending vertically, up to 60 cm. long and 10 cm. in diameter. Twining stems 6 m. long or more, sparsely beset with stellate hairs, soon glabrescent. Leaves usually opposite, occasionally alternate; petiole up to 6(–15) cm. long; blade ovate, rarely heart-shaped, with the base varying from deeply and widely cordate to subtruncate, acutely acuminate, up to 12(–14) cm. long and 7(–11) cm. broad, sparsely or densely clothed with small or longer stellate hairs, especially on the nerves, sometimes glabrescent beneath,

subglabrous or with a few scattered stellate hairs above. Inflorescences
descending, spicate. Male up to 6(–12) per leaf-axil, up to 7(–11) cm. long;
flowers sessile, directed away from the axis; perianth ultimately curving
outwards in flower, ± 3 mm. across; segments ovate, the inner narrower,
obtuse, ± 2 mm. long. Female 3(–5) per leaf-axil, up to 25 cm. long, usually
much shorter; perianth erect, almost 1·5 mm. across. Ovary densely
stellate-pubescent. Capsule as in fig. 5/5, ± 4 cm. in diameter, glabrescent.
Seeds winged all round, 2 cm. in diameter. Figs. 4, p. 17, and 5.

TANZANIA. Lushoto District: Lushoto to Shume road, Magamba Forest, 1 Mar. 1953
   (♂ fl.), *Drummond & Hemsley* 1359! & Amani, Bomole Hill, 20 Apr. 1968 (fr.),
   Renvoize 1624!; Morogoro District: Uluguru Mts., Bondwa Mt. above Morningside,
   Dec. 1953 (♀ fl. & fr.), *Eggeling* 6770!
DISTR. T3, 6; not known elsewhere
HAB. Upland rain-forest, often persisting in secondary growth after felling; 1000–
   1950 m.

SYN. *D. schimperana* Kunth var. *nigrescens* R. Knuth in E.P. IV. 43: 256 (1924).
      Types: 14 syntypes from Usambara and Uluguru Mts., including Usambara
      Mts., Dodwe Stream near Amani, *Grote* in Herb. *Amani* 3747 (EA, isosyn.)

VARIATION. The degree of indumentum of the leaves varies very considerably. Some
   specimens from the Uluguru Mts. have a few small scattered hairs on the lower
   surface, especially on the main nerves and towards the base and are soon glabrescent,
   whilst the majority are practically glabrous. In the Usambara Mts. a few are similar
   to the Uluguru Mts. plants, but many have a thick scattering of stellate hairs, especially
   beneath. One specimen, *Batty* 976 from Usambara Mts., between Oaklands and
   View Point, 17 Mar. 1970, with indumentum on its leaves resembling that of 11,
   D. schimperana, was gathered alongside a plant with considerably less dense indumentum. There are, however, so many variations in the degree of indumentum and
   no definite correlation with locality, habitat or altitude that I prefer not to recognize
   varieties by name. Considerable variation also occurs in the shape of the capsules,
   but here again I find no correlation with other characters or with the factors mentioned above. The capsules are in general considerably larger than in No. 11, *D.
   schimperana* and No. 12, *D. hirtiflora*.

NOTE. There appears to be no overlap in distribution between *D. longicuspis* and
   D. hirtiflora var. *orientalis*, as the latter species does not occur in the upland rain-
   forest where *D. longicuspis* is to be found.

14. **D. buchananii** *Benth.* in Hook., Ic. Pl. 14: 76, tt. 1397, 1398 (1882);
Bak. in F.T.A. 7: 415 (1898); R. Knuth in E.P. IV. 43: 185 (1924). Types:
Malawi, Shire Highlands, *Buchanan* 173 (♂) & 358 (♀) (both K, syn.!)

Tuber perennial, woody, irregularly shaped, often with flattened sometimes horizontal lobes, up to 30 cm. long. Plant glabrous. Twining stems
up to 9 m. long, terete. Leaves alternate; petiole up to 40(–60) cm. long;
blade very variable in shape and size, heart-shaped or triangular in outline,
entire or 2–6-lobed in the lower half (see fig. 3/4–6, p. 8), with a short
acute acumen, up to 12 cm. long and 15·5 cm. wide, often proportionally
much narrower. Inflorescences directed downwards, single in the leaf-axils.
Male raceme dense, up to 5(–6) cm. long, often much shorter; pedicels
± 2 mm. long; perianth cup-shaped, ± 10(–13) mm. across when flattened.
Stamens 6. Female inflorescences spicate, up to 35 cm. long; flowers
directed downward or somewhat spreading; perianth star-shaped, ± 8 mm.
in diameter. Ovary ± 8 mm. long. Capsule as in fig. 1/7, p. 4, up to
3·5 cm. long and 2 cm. in diameter. Seeds (fig. 1/7a) with a broad wing at
each end joined by a narrower wing, oblong, 16 mm. long, 10 mm. wide.

TANZANIA. Dodoma District: about 9 km. E. of Itigi Railway Station, 7 Apr. 1964
   (unripe fr.), *Greenway & Polhill* 11436!; Mpwapwa District: near entrance to
   Kongwa Pasture Research Station, 13 Mar. 1967 (♀ fl.), *Martin* 11!; Tunduru
   District: 96 km. west of Masasi along Tunduru road, 19 Mar. 1963 (♂ fl.), *Richards*
   17967!

DISTR. T1, 3–8; Malawi, Zambia, Rhodesia and Mozambique
HAB. Near rock outcrops in woodland and in *Combretum* thickets; 300–1350 m.

SYN. *D. buchananii* Benth. var. *ukamensis* R. Knuth in E.P. IV. 43: 185 (1924).
      Type: Tanzania, Morogoro District, Ukami, *Stuhlmann* 8283 (B, holo.)
      D. mildbraediana R. Knuth in N.B.G.B. 11: 1059 (1934). Type: Tanzania,
      Kilwa District, Mswega, *Schlieben* 2495 (B, holo.)
      D. rhacodes R. Knuth in F.R. 42: 162 (1937). Type: Tanzania, Morogoro
      District, Ukami, *Peter* 46419 (B, holo. !)

VARIATION. The plentiful material now available shows that no taxonomic value can
be placed on the leaf-shape and degree of lobing, and I have therefore placed var.
ukamensis in synonymy. Both entire and conspicuously lobed leaves may occur on
a single plant.

NOTE. *D. buchananii* has the largest flowers of any African species of *Dioscorea*.

15. **D. preussii** *Pax* in E.J. 15: 147 (1892); Bak. in F.T.A. 7: 417 (1898);
De Wild. in B.J.B.B. 4: 318 (1914); R. Knuth in E.P. IV. 43: 221 (1924);
F.W.T.A. 2: 382 (1936); Burkill in B.J.B.B. 15: 351 (1939); F.P.N.A. 3:
394 (1955); Miège in F.W.T.A., ed. 2, 3: 152 (1968). Type: Cameroun,
between Barombi and Kumba, *Preuss* 380 (B, holo.)

Tuber replaced annually, narrowly cylindric, up to 40 cm. long, with
thickened horizontal branches, up to 30 cm. long and 3·5 cm. in diameter
below. Twining stems up to 24 m. long, many-sided and narrowly winged
below, glabrescent. Leaves alternate; petiole up to 15(–25) cm. long;
blade heart-shaped, up to 12(–28) cm. long and 15(–30) cm. broad, with an
acute acumen 1(–2) cm. long; indumentum of dense to sparse woolly somewhat adpressed medifixed hairs, more dense beneath, glabrescent in age.
Inflorescences pendulous, racemose. Male up to 5 from a leaf-axil, up to
22 cm. long; pedicels ± 1 mm. long. Perianth opening flat, ± 5 mm.
across; segments slightly concave, pubescent outside, the 3 inner rather
broader. Stamens 3; staminodes 3, arching over the pistilode. Female
inflorescence spicate, axillary, up to 34 cm. long; flowers solitary, directed
downwards; perianth flat, almost 4 mm. wide. Ovary ± 8 mm. long.
Capsule as in fig. 1/8, p. 4, with a narrow wing along each suture, 4–5·5 cm.
long and 2·3 cm. in diameter. Seeds winged at both ends, ± 2·5 cm. long.

UGANDA. Acholi District: Gulu, Pagak, July 1937 (♂ fl.), *Eggeling* 3355!; Toro
District: Bwamba Forest Reserve, Mungilo, 28 July 1960 (♂ fl.), *Samwell Paulo*
610!; Mbale District: Bukedi, Terinyi, 17 Oct. 1959 (fr.), *Peal & Sakwa* 38!
DISTR. U1–4; ?T4 (see note); Senegal to Sudan, Zaire and Angola
HAB. Lowland rain-forest, woodland and secondary bushland; 650–1200 m.

SYN. *D. dawei* De Wild. in B.J.B.B. 4: 317 (1914); R. Knuth in E.P. IV. 43: 223
      (1924). Type: Uganda, Bunyoro, *Dawe* 754 (BR, holo., K, iso.!)

NOTE. Burkill (B.J.B.B. 15: 354 (1939)) states that *D. preussii* produces bulbils
[aerial tubers], but none is represented on the abundant material which I have seen;
and Dr. J. Lawton, who is familiar with the species in Nigeria, has never seen it
producing bulbils. Collections showing these bulbils would be welcome.
      A sterile specimen from Tanzania, Mpanda District, Kungwe Mts., Kahoko Hill,
17 Nov. 1963 (sterile), *Carmichael* 1028 (EA), may be this species.

16. **D. hylophila** *Harms* in P.O.A. C: 146 (1895); V.E. 2: 362, fig. 258
(1908); R. Knuth in E.P. IV. 43: 223, fig. 51 (1924); Burkill in B.J.B.B.
15: 352 (1939). Type: Tanzania, W. Usambara Mts., Lutindi, *Holst* 3423
(B, holo., K, iso. !)

Tubers replaced annually, 1–several, narrowly cylindric, up to 45 cm.
long and 15 mm. in diameter. Twining stems up to 12 m. long, terete,
glabrescent. Leaves alternate; petiole up to 14 cm. long; blade heart-

shaped, up to 23 cm. long and 22 cm. broad (usually considerably smaller), with a very acute acumen to 1(–2) cm. long; indumentum of sparse, adpressed medifixed hairs, the upper surface glabrescent. Inflorescences pendulous. Male raceme-like, axillary or fasciculate, up to 27 cm. long (often much shorter), with single pedicellate flowers or short 2–6-flowered cymules at each node; perianth cup-shaped, ± 3 mm. across; segments concave, pubescent outside, the inner rather larger. Stamens 3; staminodes 3, arching over the pistilode. Female inflorescence spicate, axillary, up to 31 cm. long; flowers solitary or with occasionally a second flower at a node, directed downwards; perianth widely cup-shaped, ± 4 mm. in diameter. Ovary ± 10 mm. long. Capsule as in fig. 1/8, p. 4, with a narrow wing along each suture, ± 6 cm. long and 3·5 cm. in diameter. Seeds (fig. 1/8a) not seen mature, probably as in 15, *D. preussii*.

TANZANIA. Lushoto District: W. Usambara Mts., road from Kwamshemshi to Sakare, 4 July 1953 (♀ fl. & young fr.), *Drummond & Hemsley* 3152!; Morogoro District: Magadu, June 1953 (♂ fl.), *Samwell Paulo* 101!; Songea District: SW. of Kitai, R. Nakawali, Mar. 1956 (♂ & ♀ fl.), *Milne-Redhead & Taylor* 9102! & 9115!
DISTR. T3, 5, 6, 8; Malawi and Zambia (Mpika)
HAB. Lowland rain-forest, riverine forest, often persisting in secondary bushland and abandoned plantations; 250–1100 m.

NOTE. The illustration published in the two German works cited above is incorrect; the capsule should be shown as ascending (not descending, as is the ♀ flower).

17. **D. dumetorum** (*Kunth*) *Pax* in E. & P. Pf. II. 5: 134 (1888); Th. Dur. & Schinz, Consp. Fl. Afr. 5: 274 (1893); Harms in P.O.A. C: 147 (1895); Bak. in F.T.A. 7: 419 (1898); V.E. 2: 359, fig. 254 (1908); De Wild. in B.J.B.B. 4: 349 (1914); F.W.T.A. 2: 382 (1936); Burkill in B.J.B.B. 15: 367 (1939); Miège in F.W.T.A., ed. 2, 3: 151 (1968); Verdc. & Trump, Common Poisonous Pl. E. Afr.: 193, figs. 19, 20/k, l (1969). Types: Ethiopia, Tigre, Djeladjeranne, *Schimper* (sect. 3) 1449 (B, isosyn.) & R. Taccaze, *Schimper* (sect. 2) 786 (B, isosyn.)

Tuber replaced annually, much divided with short or cylindric round-ended root-bearing lobes up to 2·5 cm. in diameter, spreading or descending to ± 30 cm. Twining stems up to 10 m. high, pubescent and beset sparsely or rather densely with prickles. Leaves alternate, 3-foliolate; petiole up to 20 cm. long, pubescent, usually with a few scattered prickles; leaflets with petiolule up to 10 mm. long, adpressed pubescent, glabrescent above, discolorous and lanate, rarely sparsely hirsute beneath; median leaflet obovoid, acutely acuminate, cuneate or rounded at the base, conspicuously 3-nerved from just above the base. Male inflorescence paniculate, the ultimate branches spreading in all directions forming dense subsessile cylindric spikelets up to 15 mm. long, subsessile or on peduncles up to 5 mm. long; bracts broadly ovate, adpressed to the perianth and partly concealing it, densely pubescent; perianth subglobose, glabrous, with the 3 inner segments ± 1 mm. long and the 3 outer considerably smaller and thinner. Stamens 6. Female inflorescence pendulous, spicate, 5–10 cm. long, with the flowers close together at first, the internodes elongating greatly in age; flowers directed downwards; perianth depressed subglobose, ± 2 mm. in diameter, pubescent; lobes subequal. Ovary ± 7 mm. long, densely pubescent. Capsule as fig. 2/3, p. 5, (3–)4(–5) cm. long, rather sparsely pubescent. Seeds (fig. 2/3a) ± 2 cm. long, with wing on the basal side only.

UGANDA. W. Nile District: Amua, 7 June 1936 (♂ fl.), *A. S. Thomas* 1952!; Bunyoro District: Budongo Forest, Nov. 1935 (unripe fr.), *Eggeling* 2315!; Mubende District: Kasambya, 16 May 1957 (♂ fl.), *Griffiths* 22!

KENYA. Kilifi District: Rabai, Aug. 1937 (fr.), *V. G. van Someren* in *C.M.* 7175 ! & 1886, *W. E. Taylor* !
TANZANIA. Ngara District: Rusumo ferry, 18 Mar. 1960 (fr.), *Tanner* 4795 !; Tanga District: Magunga Estate, 27 June 1952 (♂ & ♀ fl., fr.), *Faulkner* 968 !; Songea District: 32 km. E. of Songea by R. Mkurira [Mkukira], 13 Mar. 1956 (♂ & ♀ fl.), *Milne-Redhead & Taylor* 9148 ! & 9147 !; Zanzibar I., Chwaka, 5 June 1931, *Vaughan* 1938 !
DISTR. U1–4; K7; T1, 3, 4, 6–8; Z; Senegal to Ethiopia and south to Angola and Rhodesia, Mozambique and South Africa (Transvaal)
HAB. Edges of lowland rain-forest, dry evergreen forest, evergreen bushlands and on termite-hills in *Brachystegia* woodland, persisting in plantations and in secondary thickets and grasslands; 0–1650 m.

SYN. *Helmia dumetorum* Kunth, Enum. Pl. 5: 436 (1850)
    [*Dioscorea triphylla* sensu A. Rich., Tent. Fl. Abyss. 2: 316, t. 96/B (1851) *non* L.]
    D. triphylla L. var. *dumetorum* (Kunth) R. Knuth in E.P. IV. 43: 132, fig. 28 (1924)
    D. triphylla L. var. *abyssinica* R. Knuth in E.P. IV. 43: 136 (1924). Types: as *D. dumetorum*
    D. triphylla L. var. *rotundata* R. Knuth in E.P. IV. 43: 136 (1924). Type: Tanzania, Tabora District, Kakoma, *Boehm* 11a (B, holo.)

VARIATION. The varieties accepted by R. Knuth are based on leaflet shape and density of indumentum on the lower surface of the leaflets, characters which, in my opinion, are of no taxonomic significance in this species. It is of interest to note that the syntypes of var. *abyssinica* are the gatherings selected by Kunth as the types of his *Helmia dumetorum*.

NOTE. *D. dumetorum* has been slightly ennobled by cultivation, resulting in races less poisonous than the wild stock. Careful processing makes the tubers edible, and they are then eaten in times of famine.

18. **D. cochleari-apiculata** *De Wild.* in B.J.B.B. 4: 350 (1914); R. Knuth in E.P. IV. 43: 155 (1924); Burkill in B.J.B.B. 15: 370 (1939). Type: Zaire, Katanga, Lukafu, *Verdick* 267b (BR, holo.)

Tubers replaced annually, 4–5, depressed, globose, clustered, up to 10 cm. across. Twining stems up to 15 m. long, trailing down from the tops of trees, densely pubescent when young, glabrescent, beset with prickles. Leaves alternate, 3-foliolate; petiole up to 27 cm. long, pubescent, usually with a very few scattered prickles; leaflets with petiolule up to 10 mm. long, densely pubescent when young, becoming rather sparsely adpressed pubescent above in age and pubescent mainly on the nerves beneath; median leaflet obovoid to broadly obovoid, up to 24 cm. long and 29 cm. wide, acutely acuminate, cuneate or rounded at the base, conspicuously 3(–5)-nerved from just above the base. Male inflorescence paniculate, the ultimate branches spreading in all directions forming dense cylindric spike-lets up to 25 mm. long on peduncles up to 15 mm. long; bracts broadly ovate, adpressed to the perianth and partly concealing it, densely pubescent; perianth subglobose, densely pubescent, with the thin inner segments about 1 mm. long and the outer considerably smaller. Stamens 6. Female inflorescence pendulous, spicate; flowers not known. Capsule as in fig. 2/1, p. 5, 5–7 cm. long, 2·5 cm. in diameter, velutinous. Seeds (fig. 2/1a) ± 4 cm. long, with wing on basal side only.

TANZANIA. Kahama District: about 29 km. along Ushirombo road, 1 Jan. 1936 (unripe fr.), *B. D. Burtt* 5473 ! & 5473a !; Ufipa District: Lake Sundu, 23 Nov. 1960 (♂ fl.), *Richards* 13595 !; Songea District: about 6·5 km. S. of Gumbiro, 29 Mar. 1956 (unripe fr.), *Milne-Redhead & Taylor* 9375 !
DISTR. T4, 5, 7, 8; Burundi, Zaire (Katanga), Zambia and Rhodesia
HAB. Sandy lake-shores, sandy ground in *Brachystegia* woodland, termite-hills and near rock outcrops, persisting in abandoned cultivated ground; 750–1900 m.

SYN. *D. stolzii* R. Knuth in E.P. IV. 43: 136 (1924). Types: Tanzania, Mbeya
    District, Utengule, *Stolz* 2366 & Rungwe District, Bulambia, *Stolz* 1716 (both
    B, syn., K, isosyn. !)

VARIATION. In Tanzania the species is very uniform, but towards the southern end
    of its range, in Rhodesia, some plants have glabrous perianths, a character not
    observed farther north. It is possible that crosses with *D. dumetorum* occur, but
    there are no field or laboratory observations to confirm this.

NOTE. *D. cochleari-apiculata* De Wild. is a close ally of the widespread species, No. 17,
    D. dumetorum; its area of distribution is, however, relatively restricted. It is a very
    vigorous plant, and prefers well-drained sandy soils. It is strange that female in-
    florescences are not represented at Kew. The tubers of *D. cochleari-apiculata* after
    preparation are used as a famine food.

19. **D. quartiniana** *A. Rich.*, Tent. Fl. Abyss. 2: 316, t. 96/A (1851);
Th. Dur. & Schinz, Consp. Fl. Afr. 5: 275 (1893); Harms in P.O.A. C: 146
(1895); V.E. 2: 360, fig. 257 (1908); De Wild. in B.J.B.B. 4: 351 (1914);
R. Knuth in E.P. IV. 43: 151 (1924); Burkill in B.J.B.B. 15: 362 (1939);
Burkill & Perrier, Fl. Madag. 44: 20, fig. 5 (1950); Miège in F.W.T.A.,
ed. 2, 3: 151 (1968); Verdc. & Trump, Common Poisonous Pl. E. Afr.:
194 (1969). Type: Ethiopia, Tigre, Aderbati, *Quartin-Dillon* (P, holo.)

Tuber similar to that of 17, *D. dumetorum*. Twining stems up to 6 m.
high, up trees, or spreading over lower shrubs, sometimes trailing along the
ground, glabrous or sparsely pubescent with minute gland-tipped hairs and
longer simple hairs, glabrescent, without prickles. Leaves alternate with
3–5(–7) leaflets; petiole up to 11·5 mm. long, usually much shorter; leaflets
with petiolule up to 5(–15) mm. long, extremely variable in size and shape,
ovate, obovate, elliptic, lanceolate, oblanceolate or linear-lanceolate, up to
14 cm. × 11 cm., 11 cm. × 5 cm. or 10 cm. × 1 cm. according to shape,
excluding the 2–5(–9) mm. long apiculum, rounded to acute at the apex,
rounded to cuneate at the base, never 3-nerved from near the base; indumentum of sparse, short, straight or longer, denser, woolly hairs, denser
beneath, sometimes glabrescent above, usually colourless but sometimes
rusty in appearance; lateral leaflets somewhat asymmetrical. Aerial tubers
sometimes present. Male inflorescences shortly pedunculate catkins,
2–5(–7) in the axils of the upper foliage leaves, or those of very reduced
leaves, or often occurring in pendulous leafless panicles up to 40 cm. long;
peduncle 4–12(–38) mm. long; catkins up to 3(–6) cm. long; bracts usually
concave, ovate, acuminate, the apex often attenuate, 3–5(–7) mm. long,
densely pubescent towards base, otherwise glabrous, or less densely pubescent
throughout; perianth completely hidden by the bracts. Stamens 3;
staminodes 3. Female inflorescence pendulous, spicate, up to 17 cm. long,
with flowers close together at first, the internodes elongating greatly in
age; flowers directed downwards; perianth depressed subglobose, pubescent
especially on lower part of lobes. Ovary ± 5 mm. long, densely pubescent.
Capsule as fig. 2/2, p. 5, (2–)2·5(–3) cm. long, glabrescent. Seeds as fig. 2/2a.

KEY TO INFRASPECIFIC VARIANTS

Leaflets less than 4 times as long as broad:
  Leaflets densely covered beneath with weak hairs
    ± 0·5–1 mm. long, broadly ovate or obovate,
    1·2–2 times as long as broad .   .   .   . b. var. **latifolia**
  Leaflets usually subglabrous to thinly and shortly
    hairy, rarely densely covered with shorter
    stiffer hairs:

Leaflets 3–5(–7), usually broadest in the lower
two-thirds, acute to rounded at the apex,
often at least thinly hairy beneath; ♂ cat-
kins up to 3(–6) cm. long (widespread)   .   a. var. **quartiniana**
Leaflets 5–7, oblanceolate, broadest in the upper
third, rounded at the apex, glabrous except
sometimes for scattered mostly glandular
hairs on the margin and near the base
beneath; ♂ catkins up to 1·5 cm. long
(coastal) .   .   .   . d. var. **stuhlmannii**
Leaflets linear-lanceolate, 6–10 times as long as broad,
glabrous or with scattered hairs mainly beneath   c. var. **schliebenii**

a. var. **quartiniana**

Leaflets very variable in shape, ovate, elliptic, lanceolate or oblanceolate, apex acute
or obtuse, glabrous or with scattered adpressed hairs mainly beneath; ♂ catkins up to
3(–6) cm. long.

UGANDA. Kigezi District: Kachwekano, Jan. 1950 (♂ fl.), *Purseglove* 3201!; Mbale
District: Bukedi, Paya, July 1926 (♀ fl.), *Maitland* 1162!; Mubende District:
32 km. N. of Kakumiro, 30 Sept. 1957 (♂ fl. & fr.), *Peal* 6! & 5!
KENYA. Northern Frontier Province: Moyale, 25 Apr. 1952 (♂ fl.), *Gillett* 12934!;
Trans-Nzoia District: NE. slope of Elgon, Sept. 1956 (♀ fl.), *Tweedie* 1404!; Uasin
Gishu District: Ol Dane Sapuk, 27 July 1951 (♂ fl.), *Greenway* 8540!
TANZANIA. Ngara District: Nyamyaga [Nyamiaga], Mukagezi, 6 Jan. 1961 (♂ fl.),
Tanner 5595B!; Lushoto District: W. Usambara Mts., Mkusi, 10 Apr. 1953 (♀ fl.),
Drummond & Hemsley 2079!; Songea District: 32 km. E. of Songea, by R. Mkurira,
13 Mar. 1956 (♂ fl.), *Milne-Redhead & Taylor* 9149!; Zanzibar I., Chwaka, ♂ fl.,
Taylor 522!
DISTR. U1–4; K1, 3–7; T1–4, 6–8; Z; Sierra Leone east to Ethiopia and south to
Angola, Rhodesia and South Africa (Transvaal); also in Madagascar
HAB. Open spaces in upland and lowland rain-forest, riverine forest and forest edges,
termite-hills, scrub and thickets, often on hillsides and near rock outcrops, grass-
lands; (50–)300–2280 m.

SYN. *D. beccariana* Martelli, Fl. Bogos.: 83 (1886). Type: Ethiopia, Eritrea, Keren,
Mt. Deban, *Beccari* 303!
D. holstii Harms in P.O.A. C: 147 (1895). Type: Tanzania, Usambara Mts.,
Holst 527b (B, holo.)
D. ulugurensis R. Knuth in N.B.G.B. 11: 1060 (1934). Type: Tanzania,
Morogoro District, Uluguru Mts., Fisigo valley, *Schlieben* 3575 (B, holo.!)
D. peteri R. Knuth in F.R. 42: 161 (1937). Type: Tanzania, Lushoto District,
Malamba [Maramba], *Peter* 20931 (B, holo.!)
[*D. dumetorum* sensu U.K.W.F.: 722 (1974), *non* (Kunth) Pax]

b. var. **latifolia** *R. Knuth* in F.R. 42: 161 (1937). Type: Tanzania, Tabora District,
S. of Kombe, *Peter* 35425 (B, holo.!)

Leaflets broadly ovate or obovate, with a rounded apex and a filiform mucro up to
3 mm. long, densely hairy below.

TANZANIA. Mbulu District: Pienaar's Heights, 12 Mar. 1967 (young ♂ fl.), *Martin* 10!
& near Singida District boundary, Yaida valley, Endamillae gorge, 23 Jan. 1970
(♂ fl.), *Richards* 25258!; Kondoa District: about 11 km. S. of Kondoa, 17 Jan. 1962
(♂ fl.), *Polhill & Paulo* 1202!
DISTR. T2, 4, 5; also in South West Africa
HAB. Thickets, ravines and near rock-outcrops; 1100–1650 m.

SYN. *D. dinteri* Schinz in Mém. Herb. Boiss. 20: 11 (1900). Type: South West
Africa, Grootfontein, Streydfontein, *Dinter* 705 (Z, holo.)
D. quartiniana A. Rich. var. *dinteri* (Schinz) Burkill in B.J.B.B. 15: 365 (1939)

c. var. **schliebenii** (*R. Knuth*) *Burkill* in B.J.B.B. 15: 365 (1939). Type: Tanzania,
Ulanga District, Mahenge, Sali, *Schlieben* 2247 (B, holo.!)

Leaflets linear-lanceolate, 5–7, the outermost sometimes pedately arising from the next petiolule, glabrous or with scattered adpressed hairs mainly beneath; ♂ catkins up to 2·2(–3) cm. long.

TANZANIA. Uzaramo District: without locality, *Stuhlmann* 6589!; Lindi District: Lake Lutamba, 2 Feb. 1935 (♂ fl.), *Schlieben* 5943! & Rondo Plateau, Mchinjiri, Feb. 1952 (♂ fl.), *Semsei* 649!
DISTR. T6, 8; not known elsewhere
HAB. Woodland and bushland; 0–950 m.

SYN. *D. schliebenii* R. Knuth in N.B.G.B. 11: 659 (1932)

d. var. **stuhlmannii** (*Harms*) *Burkill* in B.J.B.B. 15: 365 (1939). Types: Tanzania, Bagamoyo and Uzaramo, *Stuhlmann* (B, several syntypes)

Leaflets oblanceolate with a rounded apex and short filiform mucro, 5–7, the outermost sometimes pedately arising from the next petiolule or base of adjacent leaflet, glabrous or with scattered hairs mainly towards the base beneath and on the margin; ♂ catkins up to 1·5 cm. long.

TANZANIA. Bagamoyo, Jan. 1873 (♂ & ♀ fl.), *Kirk*!; Rufiji, 3 Jan. 1931 (♂ fl.), *Musk* 164!; Lindi District: Machole, 25 May 1943 (♂ fl.), *Gillman* 1450!
DISTR. T6, 8; also in Mozambique
HAB. Coastal evergreen bushland; 0–15 m.

SYN. *D. stuhlmannii* Harms in P.O.A. C: 146 (1895)

VARIATION. *Dioscorea quartiniana* A. Rich. is the most widespread species in Africa and, having adapted itself to living outside forests, it has become extremely variable. Burkill has recognized a number of varieties, many of which are not altogether satisfactory, and I prefer to treat var. *quartiniana* as a complex at present defying normal taxonomic treatment. There are, however, three extreme forms in East Africa which not only are easily recognizable but which seem to be associated with certain ecological or edaphic conditions. Var. *latifolia* in East Africa seems to be restricted to the low rainfall central region of Tanzania and, what is of great interest, plants indistinguishable from it occur in South West Africa, a good example of a disjunct distribution. Var. *stuhlmannii* occurs in the coastal zone, a region into which var. *quartiniana* has not penetrated*, whilst var. *schliebenii* is characteristic of the SE. corner of Tanzania, a region so rich in taxa not found elsewhere in East Africa. I would add that intermediates, at least in certain characters, do occur and I consider the species would be a good subject for the application of modern taxonomic methods.

* A single specimen from Zanzibar does not belong to var. *stuhlmannii*.

INDEX TO DIOSCOREACEAE